AN ARMY OF ANTS

BY REBECCA STORM

CONTENTS

ARMY ANTS	4
ARMY ANTS UP CLOSE	6
WANDERING COLONIES	8
ANT FAMILIES	10
LIVING IN THE DARK	12
MARCHING COLONY	14
TIGHT-KNIT COMMUNITIES	16
QUEEN ANT	18
RAISING BABY ANTS	20
HUMANS AND ANTS	22
TYPES OF ANTS	24
UNIQUE ANTS	26
FUN ANT FACTS	28
GLOSSARY	30
INDEX	32

Copyright © 2025 Hungry Tomato Ltd

First published in 2025 by Hungry Tomato Ltd
F15, Old Bakery Studios, Blewetts Wharf, Malpas Road, Truro, Cornwall, TR1 1QH, UK.

No part of this publication may be reproduced, stored in a retrieval system, or transmitted in any form or by any means, electronic, mechanical, photocopying, recording, or otherwise, without prior written permission of the copyright owner.

A CIP catalogue record for this book is available from the British Library.

ISBN 9781835694183

Printed in China

Discover more at
www.hungrytomato.com

DISCLAIMER:
Insects are fascinating, but best to stay away! Don't touch or handle them – some insects can sting or get aggressive when they feel threatened.

Picture credits:
Abbreviations: m-middle, t-top, l-left, r-right, bg-background.

Alex Wild: www.myrenecos.net 4bl, 8b, 16m. FLPA: 9b, 17tr. Science photo library: George Bernard 20tr. Shutterstock: Bhupinder Bagga 19br; C. Kapados 19ml; Cornel Constantin FC,4tl, 5b; Dr. Morley Read 12ml; Fah Lerthulvanich 26b; feather collector 29tr; fercast 23br; frank 60 1bg, 17b; jirasak_kaewtongsorn 15t; josefotograf14mr; Ken Griffiths 25tl; Kevin Wells Photography 20bl; Lukas Jonaitis 7tr; mac.Rizal 13br; neslihanGorvev 9tr; NOPPHARAT718 4br; orla 28bl; Patrick K. Campbell 22b; Pavel Krasensky 10b, 18m; Pnor TKK 13t; RHJPhotos 23t; Skynetphoto 6-8bg, 31b; Tomatito 21b, 21m; Ulvur 28mr; wirestock creators 29b.

Every effort has been made to trace the copyright holders, and we apologise in advance for any unintentional omissions. We would be pleased to insert the appropriate acknowledgements in any subsequent edition of this publication.

Words in **BOLD** can be found in the glossary.

ARMY ANTS

Army ants are small **insects**. They often go on marches in very large numbers, with more than a million of them in a single group!

HOW DO THEY LIVE?

Army ants are **carnivores** – they eat other animals, mainly other bugs! They are **predators** that hunt and kill their **prey**, but they are also **scavengers** that feed on dead animals.

WHERE DO THEY LIVE?

Army ants are found in tropical forests in America and Africa. Those that live in Africa are sometimes called **driver ants**.

All ants belong to an insect group known as social insects because they live in groups called colonies.

Army ants live and work together in large colonies.

IT'S A BUG WORLD

Insects belong to a group of animals known as **arthropods**. Adult arthropods have jointed legs but do not have an inner **skeleton** made of bones. Instead, they have a tough outer "skin" called an **exoskeleton**. All insects have six legs when they are adults, and most also have at least one pair of wings.

All ants are arthropods, but not all ants have wings.

ARMY ANTS UP CLOSE

A typical army ant worker has very thin legs and no wings. This ant's exoskeleton is very tough!

An army ant is an insect, with the same parts of the body as all other adult insects. It has three parts: head, **thorax**, and **abdomen**.

The abdomen is largest part of the body and contains the stomach and other important **organs**.

The thorax is the thinnest part of the body. The ant's legs are attached here.

SIX LEG CLUB

Ants and other insects are sometimes called "hexapods" because they all have six legs ("hex" means "six" in Greek). This can be a bit confusing - all insects are hexapods, but not all hexapods are insects! For example, springtails have six legs, but they are not insects.

Springtail

The head is made up of a brain, a mouth, and a pair of **antennae**. Some army ants also have eyes, but most can't see at all.

WANDERING COLONIES

An insect colony is a single family of closely related insects. A big colony can have as many as 20 million ants, but most have between one and two million ants.

A single ant only eats a tiny amount of food each day, but if you add up all the food needed for the whole colony, it becomes a very large amount!

Army ant colonies must travel in order to find more food; if they don't, they could starve.

An ant colony searching the forest floor for prey.

However, ants are not always on the move. They have a regular stop-start **cycle** that lasts for about 35 days. There are 15 days of marching, followed by 20 days when the colony stops.

Ants marching to find food

CRAMMED WITH LIFE

The **tropical** rainforests of America and Africa are overflowing with life. There are more species of bugs in these forests than anywhere else on Earth. There is plenty of food for ants here.

A tiger moth being attacked by army ants.

ANT FAMILIES

Every colony contains up to four types of ant. Three of the types – queen, worker, and soldier – are in the colony at all times. A fourth type – the male – only appears at certain times.

Each colony has one female ant that is known as the "queen". The queen is the largest in size, and her most important job is to lay eggs.

The rest of the full-time members of the colony are female workers. Most of these workers spend all their time collecting food and carrying it back to the colony.

A queen army ant surrounded by worker ants.

Army ant soldiers are larger than the worker ants. Their role is to kill prey and defend the workers and the queen.

Worker army ants make up the majority of the colony.

FEARSOME WEAPONS

Army ant soldiers are not just bigger and stronger than the workers; they also have weapons! Their head and jaws are huge, and their main method of attack is a fearsome charge head-on. Some species of soldier ant can sting, while others can squirt a spray of deadly **venom**.

The head of a soldier army ant

LIVING IN THE DARK

Most army ant workers cannot see at all. The main ways they sense the world around them is by feeling vibrations with their legs and detecting smells with their antennae.

Army ants leading and protecting worker ants on a search for food

The antennae detect chemicals known as **pheromones** that are made by other ants in the colony.

It is the job of the few soldier ants that can see well to satisfy the colony's constant hunger. Using their sense of sight, these soldiers lead groups of ants to search for food.

When they find food, they lay a trail of pheromones back to the colony. Workers then follow the trail and collect the food while the soldiers carry on looking.

If the worker ants encounter any difficulty (such as a predator with a taste for ants!), they release alarm pheromones that quickly lead soldiers to the scene.

Worker ants take the food that the soldiers have found back to the colony.

CENTRAL CONTROL

The queen controls the activity of the whole colony through releasing special pheromones that only queens can make. Some control the soldiers, sending them out to find food, whilst other pheromomes signal to the colony when it's time to stop and rest.

Army ants stopping to rest

13

MARCHING COLONY

When ants are on the move, the whole colony marches through the forest looking like a living river. They move at a very slow speed, eating everything in their path!

FOOD PROCESSING

Although most army ants are blind, they are very efficient at collecting food. They work together to carry prey that is too heavy for a single ant to carry. They may also cut up prey with their sharp jaws and carry it back in pieces.

Army ants working together to carry an earthworm back to their colony.

When the colony stops, most ants stay still while a raiding party is sent out each day. Each raiding party can contain up to 600,000 ants!

Army ants on the move

The raiding parties are sent out in a different direction every day, until all the available food has been collected. It is then time for the colony to move on.

Army ants will eat any animals – living or dead – but the food they like best is the nests of termites and other ant species, especially their eggs! Some army ants steal the eggs of other ants and raise them as workers.

An army ant stealing eggs

TIGHT-KNIT COMMUNITIES

When they come to a stop, army ants form a very tight-knit community. They link their bodies to form a living home that can cling onto any surface.

Army ants using their bodies to build a bridge for other ants to walk across

A colony of army ants doesn't need much shelter from the weather because they can create their own. All the colony needs is a branch or a fallen log to support the combined weight of a million or so ants!

RIVER CROSSING

Ants can't swim, but that doesn't stop them crossing streams and rivers! Soldiers and workers use their bodies to create temporary "living bridges" across small streams. For larger stretches of water, the ants create "living rafts" that can float the colony across.

The nest building process starts with the strongest ants: the soldiers. They grab hold of each other to make chains. The nest is made stronger by the bodies of the worker ants that fill in all the gaps around the outside of the colony.

Inside the nest, thousands more workers link their bodies to make spaces for the queen and her eggs. The fiercest soldiers take up guard duty on the outside.

Army ants link their bodies to help them make a strong structure for a nest.

Ants create a "living bridge" to cross over water!

QUEEN ANT

The queen is the most important ant in the colony. All the others can be replaced, but there is only one queen. The main task of the queen is to lay eggs that will hatch into new workers.

A queen ant only needs to **mate** once in her adult life. The queen can then lay millions of eggs! The queen does not lay eggs while the colony is moving. This only happens when the colony rests.

A queen ant laying her eggs

When laying her eggs, the queen releases the pheromone that signals the colony to stop moving. Once the eggs have developed, the queen stops making that pheromone, and the workers and soldiers unlink their bodies, and the colony moves on.

When the colony gets too big, the queen will lay special queen and male eggs. When these eggs hatch, the males and the new queens fly away to start new colonies.

FLYING ANTS

The only army ants that have wings are queens and males. The queen ants fly away to mate and start new colonies. After a queen has mated, she chews off her wings as she doesn't need them anymore.

The males also fly away to find queens to mate with. They are much bigger than the flying queens, and are sometimes called "sausage flies". After mating, the males soon die.

A queen army ant, hatching from its egg

A pair of sausage flies

RAISING BABY ANTS

Young ants get very special care. The workers spend as much time caring for the young as they do collecting food.

This special care starts as soon as an egg is laid. Every egg has to be looked after by a worker at all times. Worker ants carry the eggs while the soldier ants guard them.

Worker army ants carrying **larvae** and eggs to a new nest

Worker ants looking after and feeding larvae

When the eggs develop into ant larvae, the workers also have to carry and care for them. The amount of work increases because the larvae have to be fed! The workers chew food for the larvae so it's easy for them to swallow.

It is only after the larvae have gone through **pupation** – turning from **pupa** into adults – that they are able to feed themselves and begin working for the colony.

INSECT DEVELOPMENT

Insects develop from eggs in two different ways. With many types of insect, including ants, bees, and beetles, the eggs hatch into larvae that look very different from the adults.

However, with other insects, such as cockroaches and grasshoppers, the eggs hatch into **nymphs** that already have the adult body shape.

Egg   Larvae   Pupa

Growth of an ant from pupa to adult

HUMANS AND ANTS

The idea of a whole colony of army ants may seem terrifying! However, despite their scary reputation, army ants are not really harmful to humans.

Even when walking at a slow pace, humans can move much faster than a colony of army ants. So, people who live or work in tropical rainforests have no difficulty getting out of their way.

A close-up of a Burchell's army ant

Many people who live in tropical forests welcome the arrival of army ants. Not only do the ants eat any insect pests living there, but they also eat any dead insects that have collected in corners and underneath furniture. A visit from army ants is like a "spring cleaning" session.

Ants inside a house

INTERACTING WITH OTHER ANIMALS

The arrival of army ants is good news for some animals, too. As the ants get closer, many large insects, such as cockroaches and beetles, come out of hiding to try and escape. These insects may be fast enough to outrun the ants, but they are likely to be spotted by larger predators! Some birds follow army ants to catch the large insects that are driven out by them.

A magpie with a beetle in its beak

TYPES OF ANTS

While army ants are constantly on the move, other woodland ant species follow a more relaxed lifestyle.

WOOD ANTS

Wood ants are so-called as they live in woods and forests in places with a warm climate. Colonies of wood ants build their nests completely underground out of fallen leaves and woodland scraps. In cooler regions, they sometimes construct a mound above the nest. These mounds can get very big indeed!

A wood ant nest

Leafcutter ants

LEAFCUTTER ANTS

Leafcutter ants cut out pieces of leaf and carry them back to their underground nest. The ants line their tunnels with these bits of leaf and allow a fungus to grow on them. This fungus provides these ants with a steady supply of food.

ACACIA ANTS

In areas that are too dry for thick forests, acacia ants build their nests at the base of thorny trees. The ants feed on **nectar** made by the tree and keep the tree clear of other insects. If any other animal tries to eat the tree, the ants rush to bite and sting them.

Acacia ants on an acacia tree

UNIQUE ANTS

There are about 10,000 different species of ant. Most of them are instantly recognisable as ants, although a few are a lot more unusual...

REPLETE ANTS

Bees are not the only insects that make and store **honey**! Some ant species, including honeypot ants, collect nectar from flowers and feed it to special workers that are known as **repletes**. Replete ants store this excess food and then **regurgitate** it for other ants to eat. A replete ant can swell up to the size of a grape!

FIRE ANTS

These red-coloured ants were once only found in the hot, wet forests of South America, but they have accidentally been introduced to other countries and have become a serious pest. Fire ants have a sting that injects a very strong venom that can kill young farm animals.

VELVET ANTS

Despite its name, this "ant" is in fact a species of wasp! They sneak into bumblebee colonies to lay their own eggs on their larvae. When the wasp eggs hatch, the wasp larvae eat the bee larvae.

FUN ANT FACTS

There's so much more to know about ants! Delve into some fantastic facts about this incredible, tiny species.

A COLONY OF ARMY ANTS CAN...

kill and eat up to 100,000 animals (mostly insects) in a day!

IF YOU GATHERED UP...

all the humans in the world and all the ants, they would weigh the same!

ARMY ANT SPECIES...

existed at the time of dinosaurs, 100 million years ago, and have not changed since then!

AFTER A MALE ARMY ANT...

mates with the queen, he dies within 48 hours.

ALTHOUGH ARMY ANTS HAVE A...

painful sting, they don't come into contact with humans very often.

IN SOME SPECIES OF...

African army ant (driver ants), the queen produces eggs daily. She can make up to four million a month..

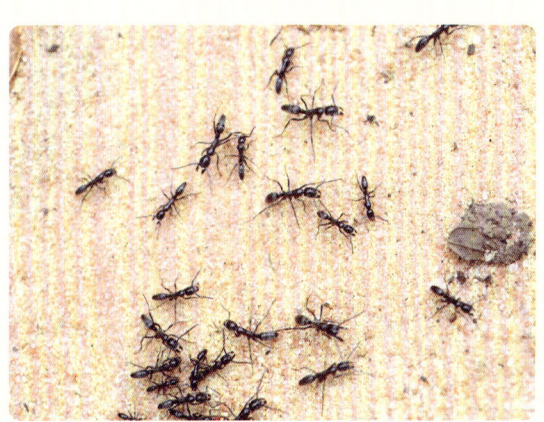

ARMY ANTS CAN CLIMB UP...

trees, and eat the birds and lizards that live there!

ARMY ANTS HAVE BEEN KNOWN TO...

kill and eat animals as big as horses!

WORKERS LIVE FOR...

several months, while the queen may live up to several years.

GLOSSARY

Abdomen – the largest part of an insect's three-part body: the abdomen, contains most of the important organs.

Antennae – a pair of special sense organs found at the front of the head on most insects.

Arthropod – any bug that has jointed legs; insects and spiders are arthropods.

Carnivores – animals that eat meat.

Colony – a group of insects, or other living things, which live very closely together.

Cycle – a series of repeated events that follow a regular pattern.

Driver ants – army ants that live in Africa.

Exoskeleton – a hard outer covering that protects and supports the bodies of some animals.

Honey – sweet, syrupy substance produced by honeybees and some ants.

Insect – a type of very small animal with six legs, a body divided into three parts, and usually two pairs of wings.

Larvae – a group of wormlike creatures called larva that are at the juvenile (young) stage in the life cycle of insects.

Mate – one of a pair of animals that live or have babies together.

Nectar – a sweet, sugary substance produced by flowering plants and used by honeybees and some ants to make honey.

Nymphs – the juvenile (young) stage in the life cycle of insects that do not produce larvae.

Organ – a part of an animal's body that performs a particular task, e.g. the heart pumps blood.

Pheromones – types of chemicals that are released by animals to communicate with each other.

Predators – animals that hunt and eat other animals.

Prey – an animal that is eaten by other animals.

Pupa – An insect larva that is in the process of turning into an adult.

Pupation – the process by which insect larvae change their body shape to the adult form.

Queen – the largest army ant in a colony; the queen is the only female army ant that can lay eggs.

Regurgitate – to bring up partially digested food.

Repletes – a special group of ant workers that store honey inside their bodies.

Scavengers – animals that eat dead and rotting plants and animals.

Skeleton – an internal structure of bones that support the bodies of large animals such as mammals, reptiles, and fish.

Soldier – a female army ant that is larger than a worker.

Thorax – the middle part of an insect's body where the legs are attached.

Tropical – belonging to the region around the Earth's equator where the climate is always hot.

Venom – a poison some animals use to hurt or kill their prey.

Worker – a female army ant, nearly all the army ants in a colony are workers.

INDEX

A
abdomen 6–7, 30
acacia ants 25
adults 5–6, 21, 28
Africa 4, 29
animal tissue 15, 22–23
antennae 6, 12, 30
ants, different species 24–27
aphids 25
Argentine ant 27
arthropods 5, 30

B
bees 27
behaviours 24–25
birds 23, 29
blindness 6, 12–13, 15
blood loss 23
bridge-building 16–17

C
carnivores 4, 30
caterpillars 30
Central America 4, 8–9, 29
cocoons 28
colonies 5, 8–11, 30
cycles 9, 30

D
driver ants 4, 29, 30

E
eggs 15, 17–20, 28–29
exoskeleton 5, 30
eyes 6, 11, 12–13

F
family groups 5, 8–11
females 10–11, 18–19
fire ants 27
flying ants 19
food 8–9
forests 4, 9
fungi 25

G
garden ants 25

H
head 6, 11, 12–13
hexapods 7
honey 26, 30
honeydew 25
humans 22–23, 29

I
insect 4-5, 6–7, 8, 20–21, 22–23,
insect development 21

J
jaws 11, 15

L
larvae 14–15, 20–21, 27–28, 30
leafcutter ants 25
legs 5, 7, 12
life cycle 28–29
linked bodies 16–17
liquid food 29
living bridges/rafts 16–17

M
males 10, 19, 29
mating 18–19, 29
mounds 24

N
nectar 25–26, 30
nests 16–17, 19, 24
nymphs 21, 30

O
organs 7, 31

P
pavement ants 14
pests 27
pheromones 12–13, 15, 19
poison 11, 27
predators 4, 11, 13, 23, 31
prey 4, 8–9, 11, 14–15, 22–23, 31
pupae 21, 28, 31
pupation 21, 31

Q
queen ants 10–11, 13, 17–19, 29, 31

R
raiding parties 14–15
regurgitation 26, 31
replete ants 26, 31
rivers 17

S
sausage flies 19
scavengers 4, 31
scouts 12–13
skeleton 5, 31
social insects 5
soldier ants 4, 10–13, 17, 20, 31
springtails 7
stings 11, 25, 27, 29
stop-start cycle 9

T
termites 15
thorax 6–7, 31
tropical regions 4, 9, 22–23, 31

V
velvet ant 27
vibrations 12

W
wasps 27
water 17
wings 5, 19
wood ants 24
workers 31